GETTING READY FOR GEOMETRY

20 DIFFERENT SKILLS AND TOPICS THAT STUDENTS SHOULD BE PROFICIENT IN BEFORE ENTERING GEOMETRY

TABLE OF CONTENTS

PAGE	TOPIC
1	THE NUMBER PROPERTIES
2	CALCULATING SLOPE
3	GRAPHING IN SLOPE-INTERCEPT FORM
4	DETERMINING PARALLEL AND PERPENDICULAR LINES
5	SOLVING MULTI-STEP EQUATIONS
6	SOLVING INEQUALITIES
7	SUBSTITUTION TO SOLVE SYSTEMS OF EQUATIONS
8	FACTORING TRINOMIALS
9	SIMPLIFYING RADICALS
10	OPERATIONS WITH RADICALS
11	CLASSIFYING SEGMENTS, RAYS, and LINES
12	NUMBER OF EDGES & VERTICES
13	ANGLE MEASUREMENTS
14	TYPES OF TRIANGLES
15	TYPES OF QUADRILATERALS
16	PARTS OF A CIRCLE
17	AREA FORMULAS OF BASIC SHAPES
18	VOLUME FORMULAS OF BASIC FIGURES
19	BASIC TRANSFORMATIONS
20	CONGRUENT OR SIMILAR

@iteachalgebra

THE NUMBER PROPERTIES

Match each expression with the property that it shows.

$5 + 0 = 5$	Commutative Property of Addition
$5(1) = 5$	Associative Property of Addition
$5(0) = 0$	Additive Identity
$2 + 3 = 3 + 2$	Distributive Property
$2(3) = 3(2)$	Commutative Property of Multiplication
$2 + (3 + 4) = (2 + 3) + 4$	Associative Property of Multiplication
$2(3 \cdot 4) = (2 \cdot 3)4$	Zero Product Property
$3(2 + 5) = 6 + 15$	Multiplicative Identity

@iteachalgebra

CALCULATING SLOPE

Find the slope between the given points or on the graph.

(1, 3) and (5, 8)   (-2, 7) and (5, 4)   (1, -3) and (0, 8)

(-1, -9) and (4, 0)   (-8, 8) and (-2, 8)   (-4, 9) and (-4, -8)

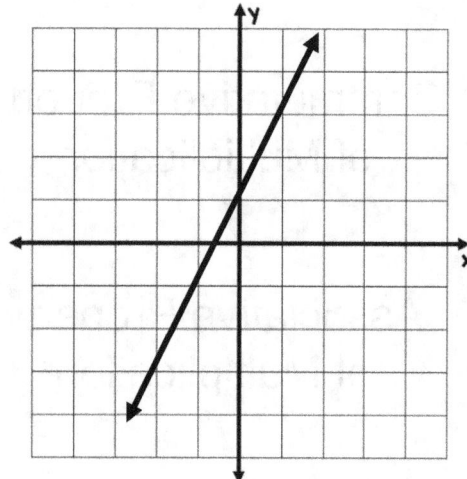
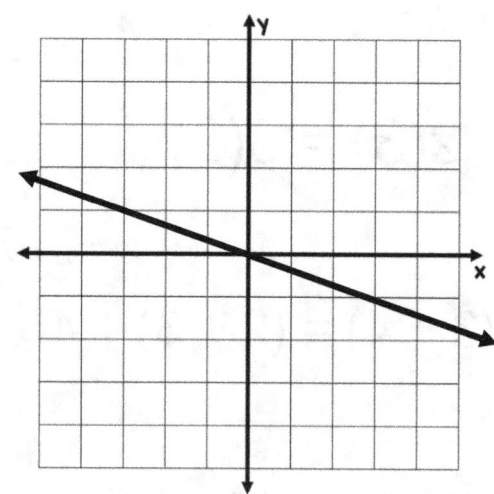

GRAPHING IN SLOPE-INTERCEPT FORM

y = x + 3

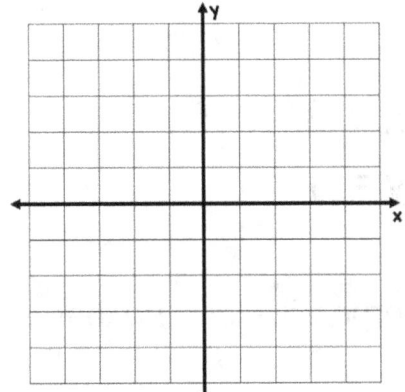

y = x - 1

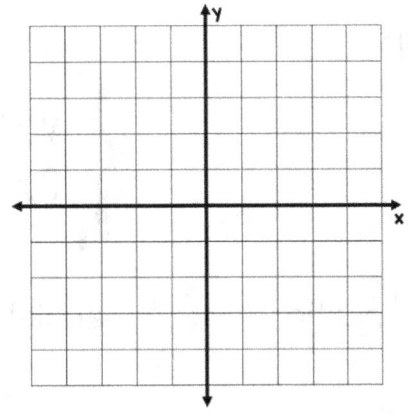

y = 2x + 3

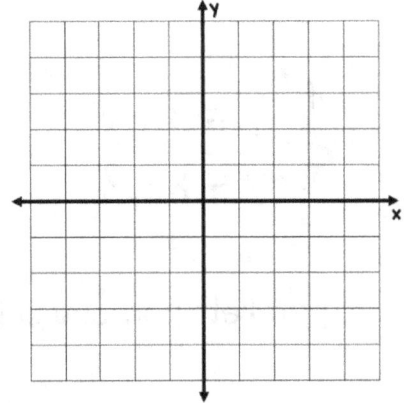

y = -2x

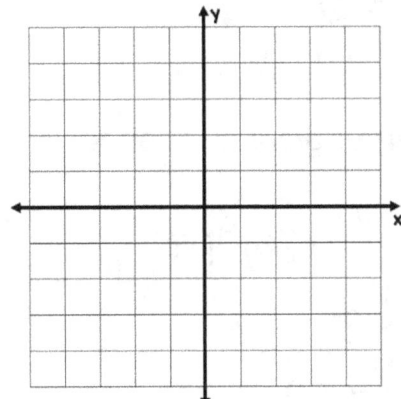

y = -x + 3

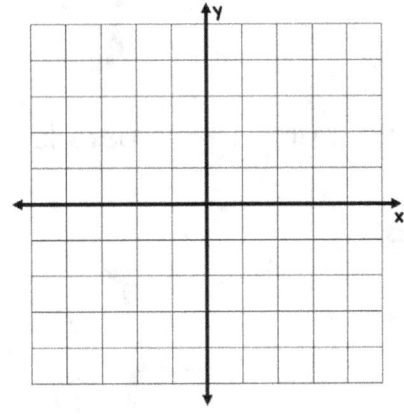

y = -3x + 3

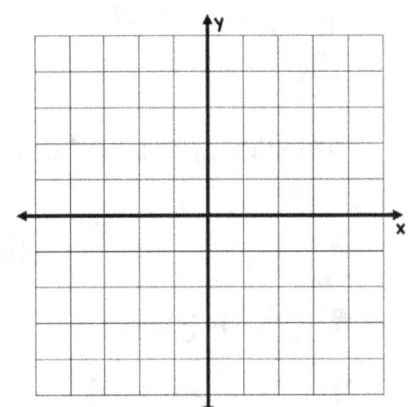

$y = \frac{1}{2}x - 4$

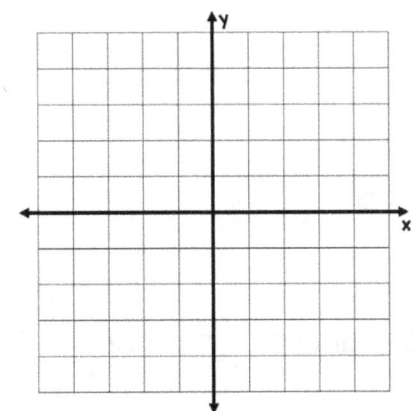

$y = -\frac{3}{2}x + 1$

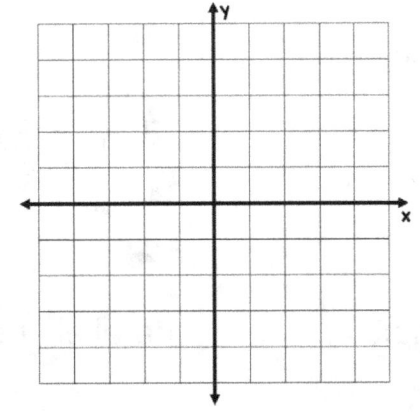

$y = \frac{4}{3}x - 3$

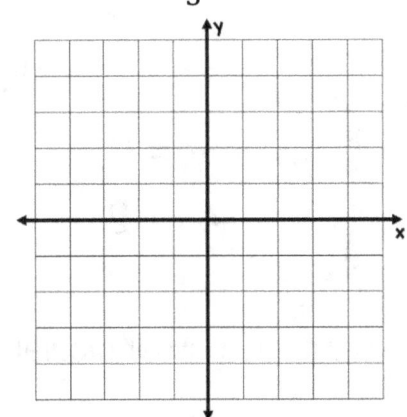

@iteachalgebra

PARALLEL & PERPENDICULAR

Circle whether each pair of equations is parallel, perpendicular, or neither.

slope:
$$\begin{cases} y = x + 3 \\ y = x - 2 \end{cases}$$
parallel   perpendicular   neither

slope:
$$\begin{cases} y = 2x + 3 \\ 2x - y = 4 \end{cases}$$
parallel   perpendicular   neither

slope:
$$\begin{cases} y = -x \\ y = x + 4 \end{cases}$$
parallel   perpendicular   neither

slope:
$$\begin{cases} y = 3x + 3 \\ x - 3y = 9 \end{cases}$$
parallel   perpendicular   neither

slope:
$$\begin{cases} 2x + 3y = 6 \\ 3x - 2y = 4 \end{cases}$$
parallel   perpendicular   neither

slope:
$$\begin{cases} y = \frac{2}{5}x + 3 \\ 2x - 5y = 10 \end{cases}$$
parallel   perpendicular   neither

slope:
$$\begin{cases} 4x + y = 6 \\ y = -4x - 2 \end{cases}$$
parallel   perpendicular   neither

slope:
$$\begin{cases} y = 5x + 3 \\ x + 4y = 8 \end{cases}$$
parallel   perpendicular   neither

@iteachalgebra

SOLVING MULTI-STEP EQUATIONS

Solve each equation. Simplify your answer.

$3(x + 4) = 2.5(x - 6)$

$2(x - 5) + 7 = -3(2x - 6)$

$\frac{1}{2}(4x - 8) = \frac{3}{4}(8x + 4)$

$\frac{1}{2}x + 5 = \frac{2}{5}x - 8$

$\frac{2}{3}(5x + 6) = \frac{3}{2}(8x - 4)$

$\frac{1}{3}x + \frac{1}{4} = \frac{2}{3}x - \frac{1}{6}$

@iteachalgebra

SOLVING INEQUALITIES

Solve the inequalities.

$30 + 2x < 17$ $15 < -4x + 18$ $6 \leq 4x + 80$

$10 - 2x \leq 17$ $-12 > -3x - 12$ $-9 \leq -5x - 33$

$8 + 2x < -x + 17$ $4x - 9 \leq 5x + 80$

$5 - 2x \geq 6(x - 3)$ $-3(3 + x) \leq -6x - 11$

@iteachalgebra

SUBSTITUTION TO SOLVE SYSTEMS

Solve each system by substitution.

$$\begin{cases} y = -2x \\ y = x + 3 \end{cases}$$

$$\begin{cases} y = 3x + 3 \\ x - 3y = 9 \end{cases}$$

$$\begin{cases} 2x + y = 6 \\ x = 2y - 1 \end{cases}$$

$$\begin{cases} y = \frac{2}{5}x + 3 \\ 2x - 5y = 10 \end{cases}$$

$$\begin{cases} x = -4 \\ y = 5 \end{cases}$$

$$\begin{cases} 2x + 3y = 6 \\ y = -3x - 1 \end{cases}$$

@iteachalgebra

FACTORING TRINOMIALS

Factor each trinomial.

$x^2 + 5x + 4$ $x^2 + 8x + 16$ $x^2 - 6x + 8$

$x^2 - 6x - 7$ $x^2 + 5x + 6$ $x^2 - 10x + 25$

$2x^2 + 7x + 3$ $3x^2 - 13x + 4$ $5x^2 + 7x - 6$

Solve the polynomial equation.

$x^2 + 9x = -8$ $2x^2 = 7x - 3$ $3x^2 + 15x = -18$

@iteachalgebra

SIMPLIFYING RADICALS

Simplify each radical expression.

$\sqrt{4}$ $\sqrt{6}$ $\sqrt{8}$ $\sqrt{9}$ $\sqrt{10}$

$\sqrt{12}$ $\sqrt{18}$ $\sqrt{25}$ $\sqrt{28}$ $\sqrt{32}$

$\sqrt{40}$ $\sqrt{48}$ $\sqrt{50}$ $\sqrt{55}$ $\sqrt{60}$

$\sqrt{64}$ $\sqrt{72}$ $\sqrt{90}$ $\sqrt{99}$ $\sqrt{120}$

$\sqrt{150}$ $\sqrt{160}$ $\sqrt{200}$ $\sqrt{256}$ $\sqrt{300}$

@iteachalgebra

OPERATIONS WITH RADICALS

Simplify each radical expression.

$\sqrt{2} + \sqrt{2}$ $\qquad$ $4\sqrt{3} + \sqrt{3}$ $\qquad$ $5\sqrt{6} + 2\sqrt{6}$

$\sqrt{2} - \sqrt{2}$ $\qquad$ $4\sqrt{3} - \sqrt{3}$ $\qquad$ $5\sqrt{6} - 2\sqrt{6}$

$\sqrt{2} \cdot \sqrt{2}$ $\qquad$ $4\sqrt{3} \cdot \sqrt{3}$ $\qquad$ $5\sqrt{6} \cdot 2\sqrt{6}$

$\sqrt{72} + \sqrt{50}$ $\qquad$ $4\sqrt{45} - \sqrt{125}$ $\qquad$ $5\sqrt{27} + 2\sqrt{5}$

@iteachalgebra

CLASSIFYING SEGMENTS, RAYS, & LINES

Determine the segments, rays, and lines from the diagram.

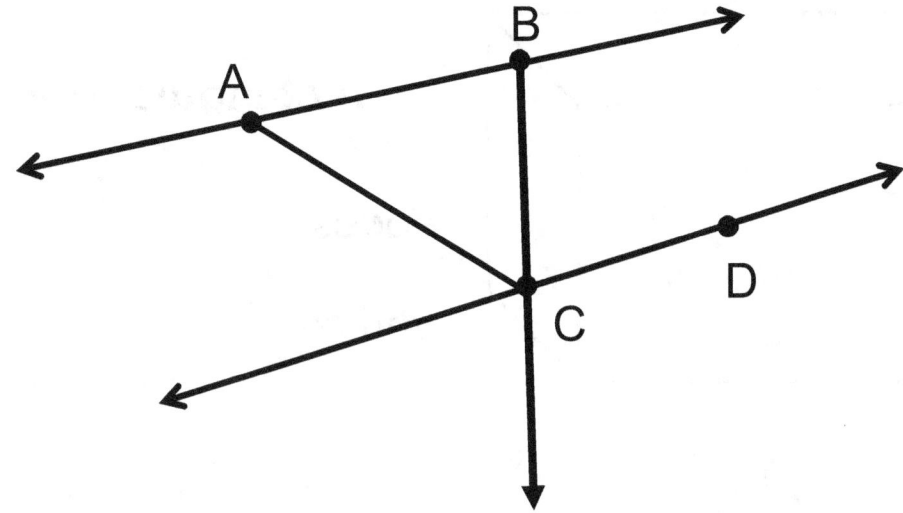

SEGMENTS	RAYS	LINES

Determine whether each statement is true or false.

Two lines can intersect at exactly one point.	
Two lines can intersect at exactly two points.	
The are an infinite number of points on a line.	
A ray has an arrow at one end.	
A segment and a line are identical.	

@iteachalgebra

NUMBER OF EDGES & VERTICES

12

List the number of edges and vertices for each figure.

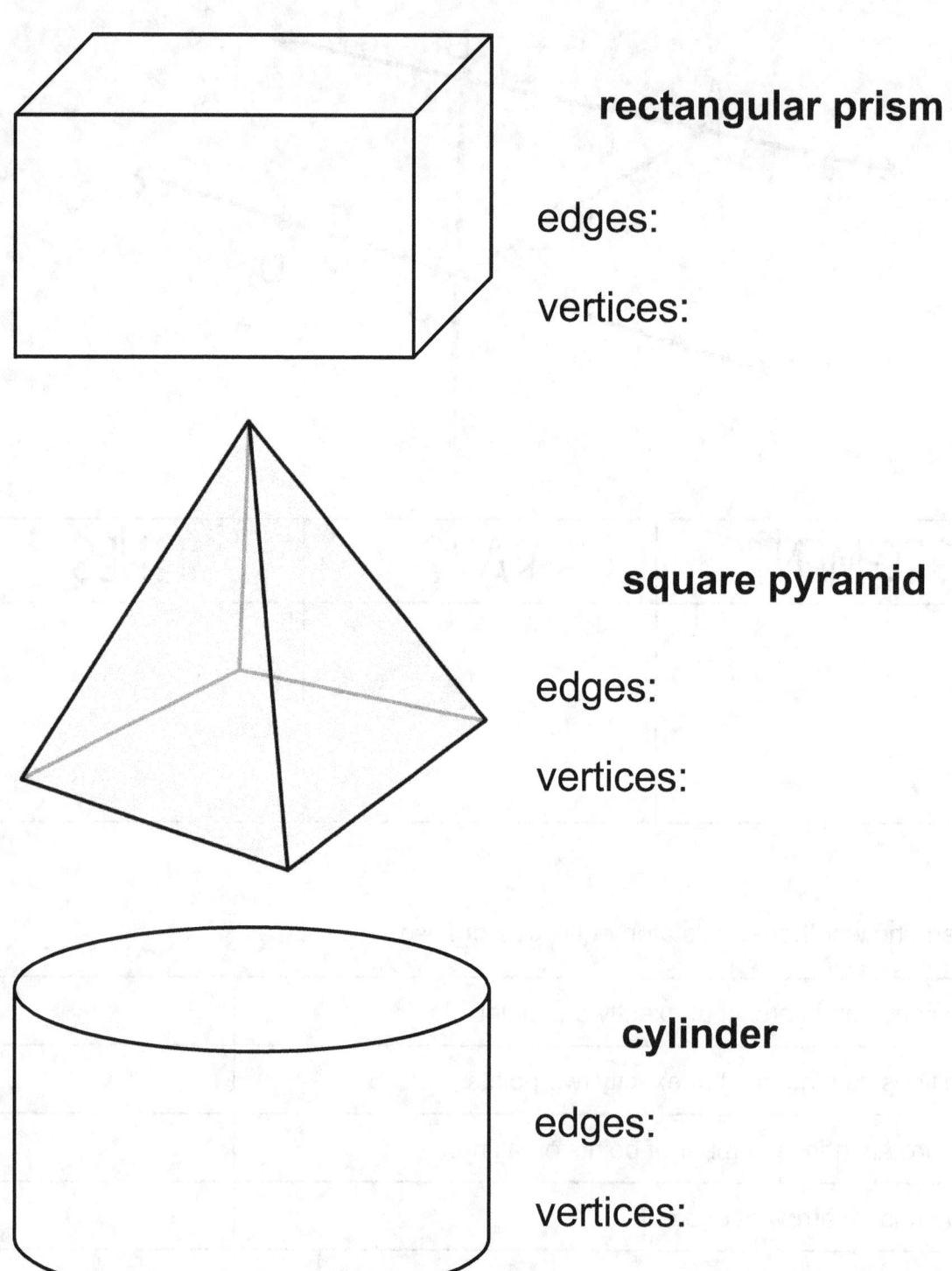

rectangular prism

edges:

vertices:

square pyramid

edges:

vertices:

cylinder

edges:

vertices:

@iteachalgebra

ANGLE MEASUREMENTS

Circle the type of angle shown and the best approximate measure of the angle.

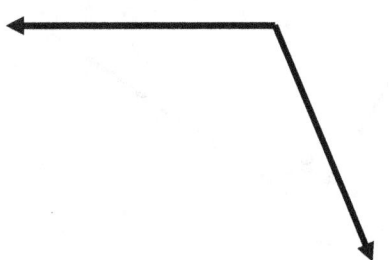

acute   60

obtuse   100

right   90

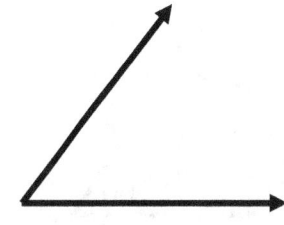

acute   60

obtuse   100

right   90

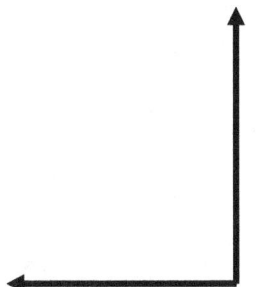

acute   60

obtuse   100

right   90

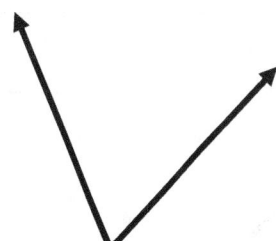

acute   60

obtuse   100

right   90

@iteachalgebra

TYPES OF TRIANGLES

14

Name the triangle based on its sides and angles.
Names include equilateral, isosceles, and scalene, acute, obtuse, and right.

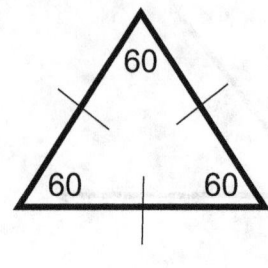

_____

_____

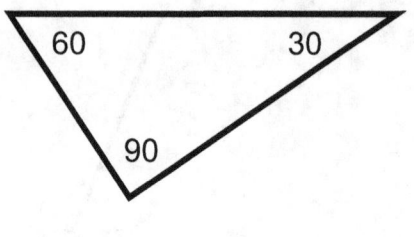

_____

_____

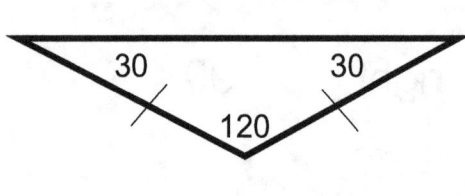

_____

_____

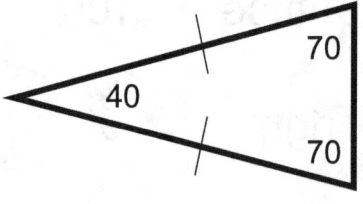

_____

_____

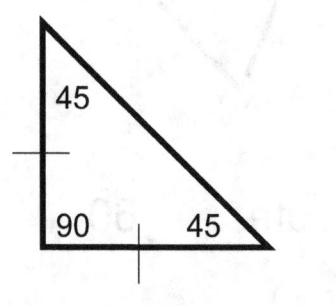

_____

_____

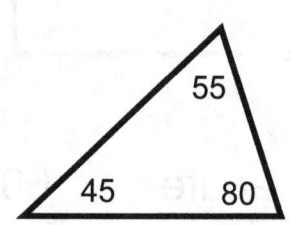

_____

_____

@iteachalgebra

TYPES OF QUADRILATERALS

Determine if the quadrilateral is a square, rectangle, rhombus, trapezoid, isosceles trapezoid, parallelogram, or more than one of those names.

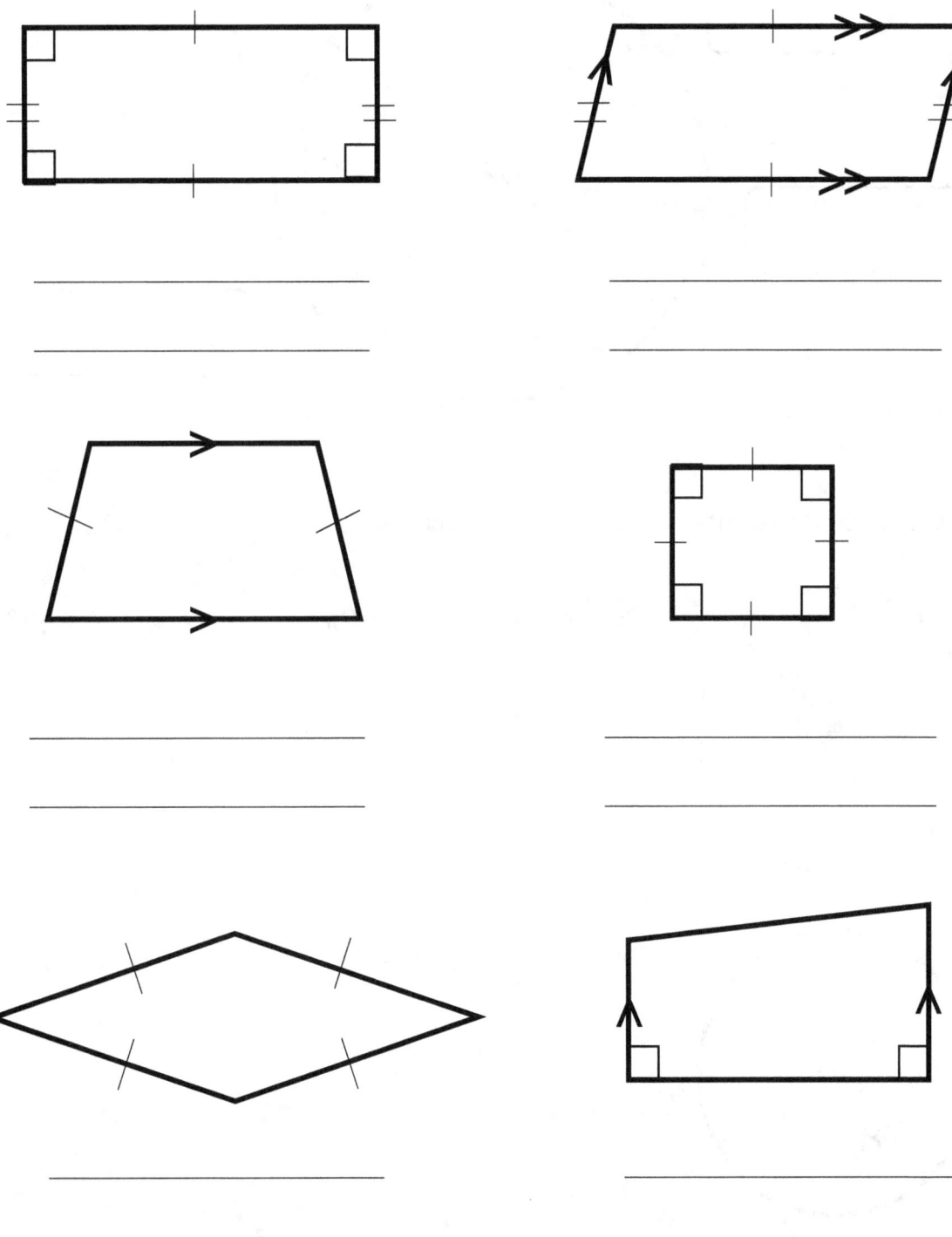

PARTS OF A CIRCLE

Given the circle, name each part.

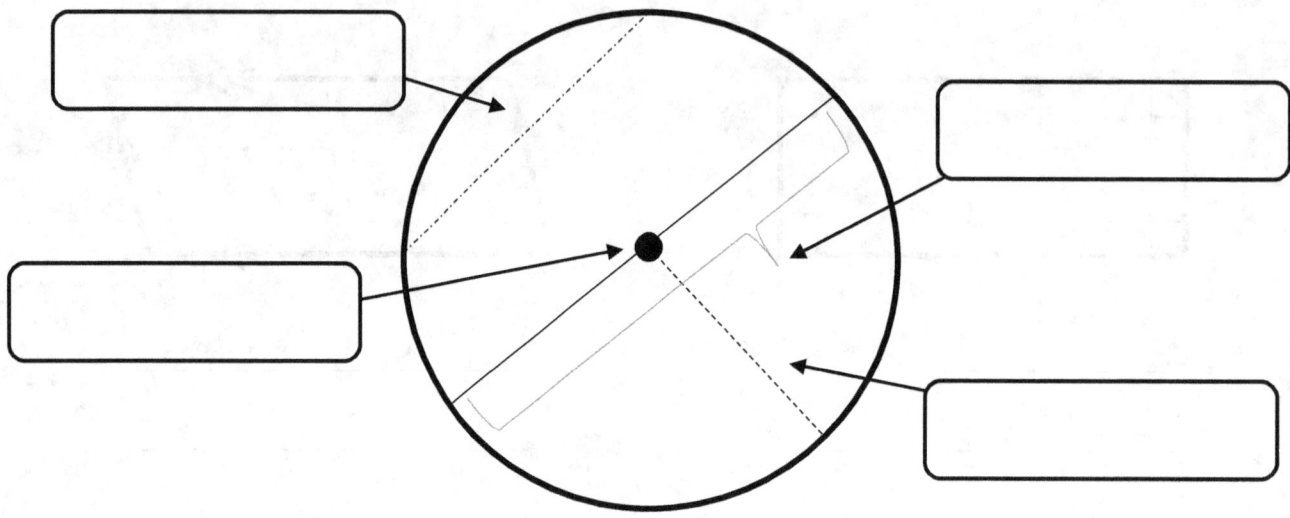

Find the circumference and area of each circle.

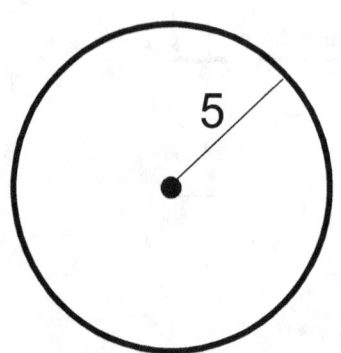

Circumference: $C = 2\pi r$ Area: $A = \pi r^2$

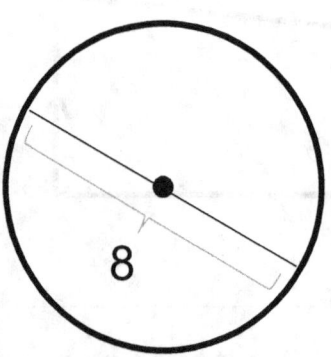

Circumference: $C = \pi d$ Area: $A = \pi r^2$

@iteachalgebra

AREA FORMULAS

Calculate the area of each figure.

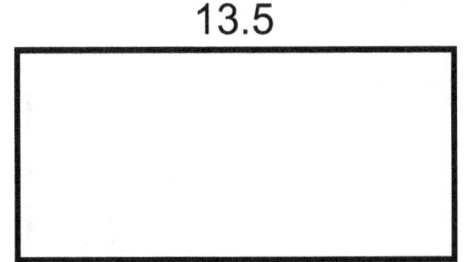

rectangle
$A = lw$

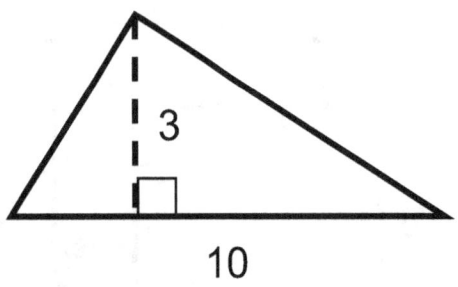

triangle
$A = \frac{1}{2}bh$

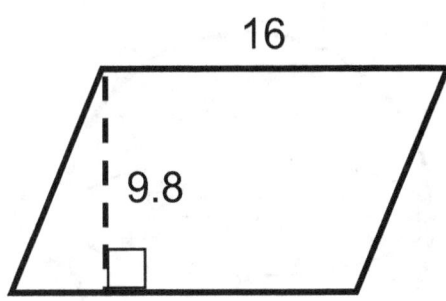

parallelogram
$A = bh$

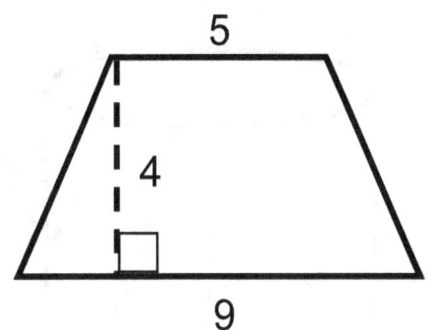

trapezoid
$A = \frac{1}{2}h(b_1 + b_2)$

@iteachalgebra

VOLUME FORMULAS

Calculate the volume of each figure.

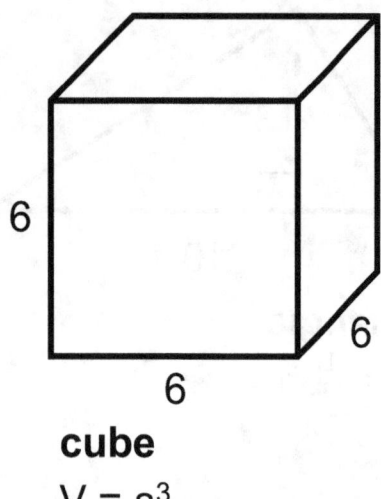

cube
$V = s^3$

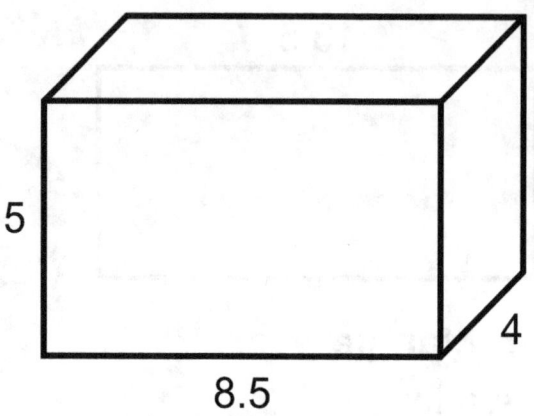

rectangular prism
$V = lwh$

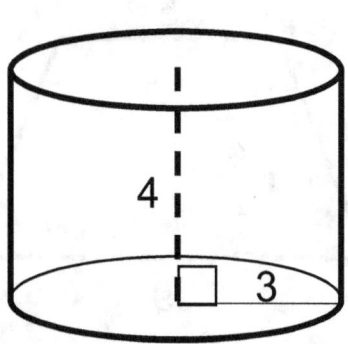

cylinder
$V = \pi r^2 h$

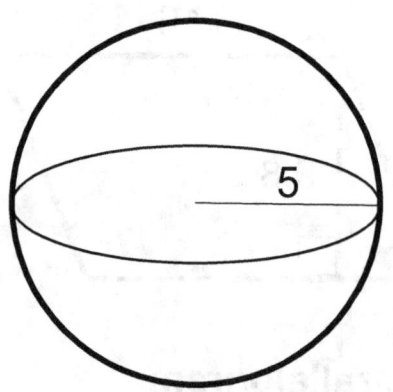

sphere
$V = \frac{4}{3}\pi r^3$

@iteachalgebra

TRANSFORMATIONS

19

Determine the type of transformation shown in each diagram as a translation, rotation, reflection, or dilation.

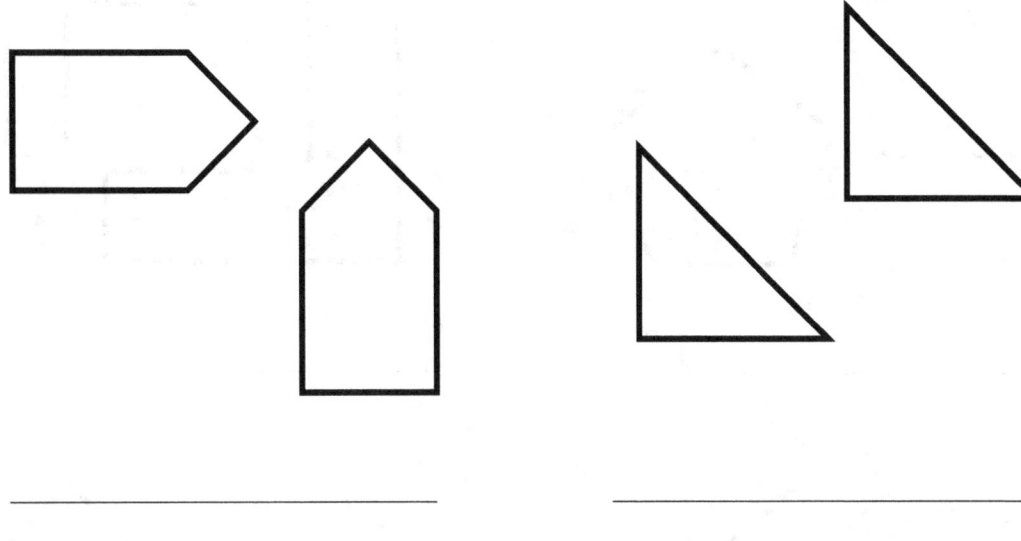

_____   _____

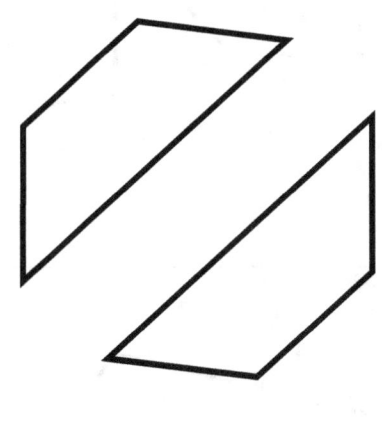   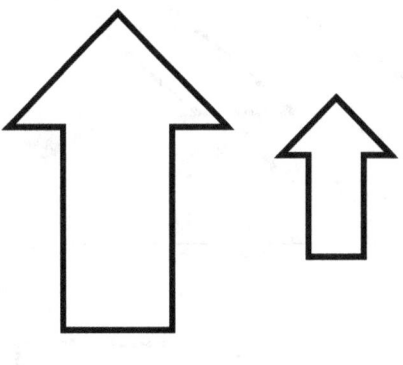

_____   _____

@iteachalgebra

CONGRUENT OR SIMILAR

Determine whether the figures shown are congruent or similar.

ANSWER KEY

ANSWER KEY

THE NUMBER PROPERTIES

Match each expression with the property that it shows.

5 + 0 = 5 — Additive Identity

5(1) = 5 — Multiplicative Identity

5(0) = 0 — Zero Product Property

2 + 3 = 3 + 2 — Commutative Property of Addition

2(3) = 3(2) — Commutative Property of Multiplication

2 + (3 + 4) = (2 + 3) + 4 — Associative Property of Addition

2(3·4) = (2·3)4 — Associative Property of Multiplication

3(2 + 5) = 6 + 15 — Distributive Property

@iteachalgebra

ANSWER KEY

CALCULATING SLOPE

Find the slope between the given points or on the graph.

(1, 3) and (5, 8)

$$m = \frac{8-3}{5-1} = \frac{5}{4}$$

(-2, 7) and (5, 4)

$$m = \frac{4-7}{5-(-2)} = \frac{-3}{7}$$

(1, -3) and (0, 8)

$$m = \frac{8-(-3)}{0-1} = \frac{11}{-1}$$

$$m = -11$$

(-1, -9) and (4, 0)

$$m = \frac{0-(-9)}{4-(-1)} = \frac{9}{5}$$

(-8, 8) and (-2, 8)

$$m = \frac{8-8}{-2-(-8)} = \frac{0}{6}$$

$$m = 0$$

(-4, 9) and (-4, -8)

$$m = \frac{-8-9}{-4-(-4)} = \frac{-17}{0}$$

$$m = \text{undefined}$$

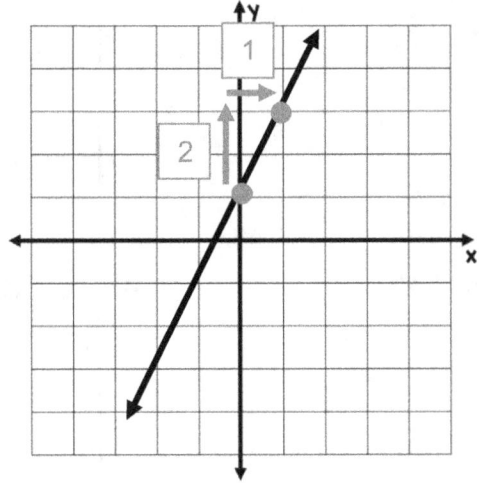

$$\frac{rise}{run} = \frac{2}{1} = 2$$

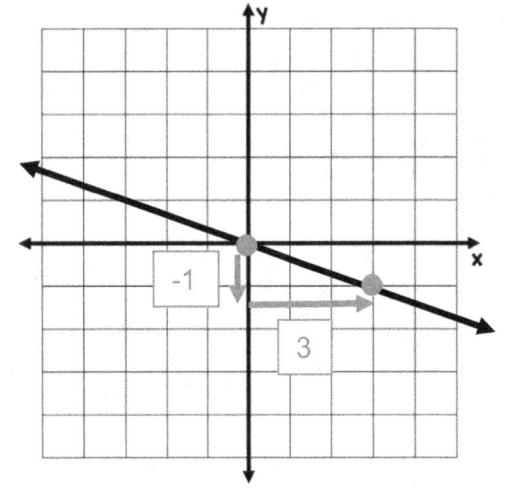

$$\frac{rise}{run} = \frac{-1}{3}$$

@iteachalgebra

ANSWER KEY

GRAPHING IN SLOPE-INTERCEPT FORM

$y = x + 3$

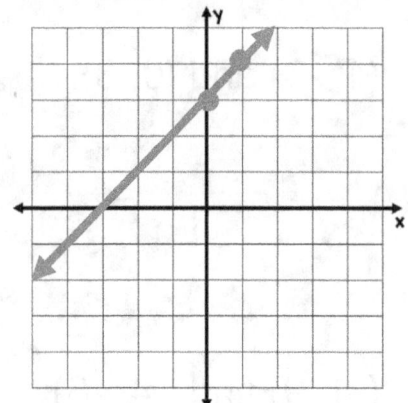

$y = x - 1$

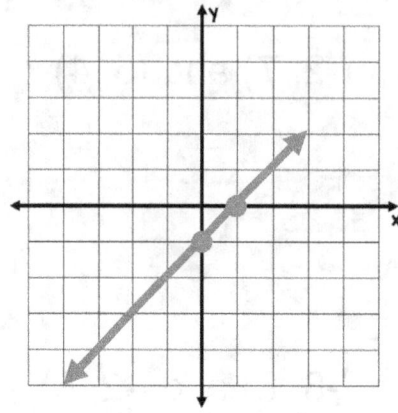

$y = 2x + 3$

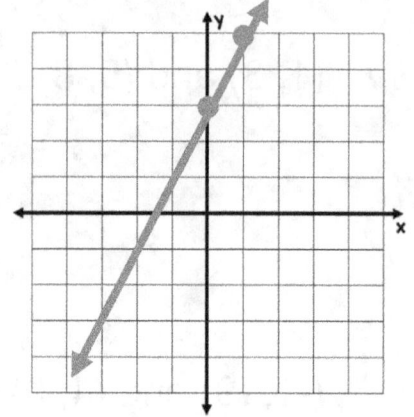

$y = -2x$

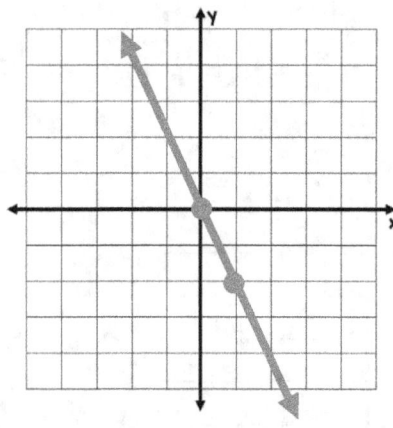

$y = -x + 3$

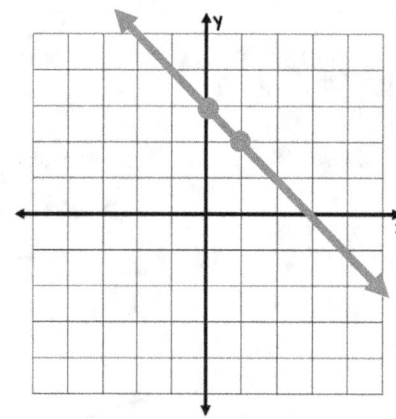

$y = -3x + 3$

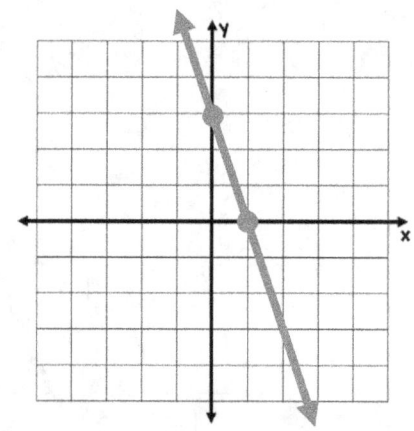

$y = \frac{1}{2}x - 4$

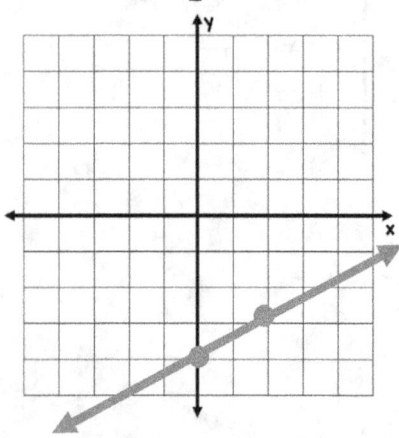

$y = -\frac{3}{2}x + 1$

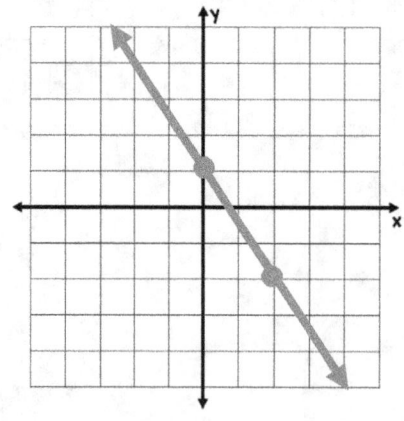

$y = \frac{4}{3}x - 3$

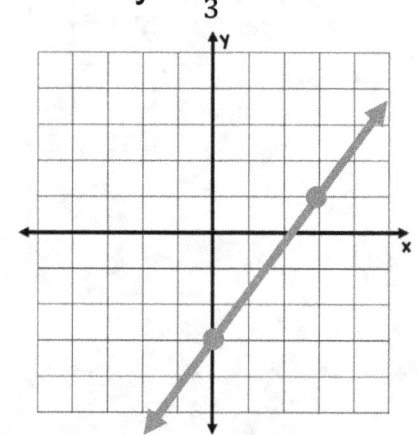

@iteachalgebra

ANSWER KEY

PARALLEL & PERPENDICULAR

Circle whether each pair of equations is parallel, perpendicular, or neither.

slope:
$\begin{cases} y = x + 3 \\ y = x - 2 \end{cases}$ $m = 1$
$m = 1$

(parallel)   perpendicular   neither

slope:
$\begin{cases} y = 2x + 3 \\ 2x - y = 4 \end{cases}$ $m = 2$
$m = 2$

(parallel)   perpendicular   neither

slope:
$\begin{cases} y = -x \\ y = x + 4 \end{cases}$ $m = -1$
$m = 1$

parallel   (perpendicular)   neither

slope:
$\begin{cases} y = 3x + 3 \\ x - 3y = 9 \end{cases}$ $m = 3$
$m = \frac{1}{3}$

parallel   perpendicular   (neither)

slope:
$\begin{cases} 2x + 3y = 6 \\ 3x - 2y = 4 \end{cases}$ $m = -\frac{2}{3}$
$m = \frac{3}{2}$

parallel   (perpendicular)   neither

slope:
$\begin{cases} y = \frac{2}{5}x + 3 \\ 2x - 5y = 10 \end{cases}$ $m = \frac{2}{5}$
$m = \frac{2}{5}$

(parallel)   perpendicular   neither

slope:
$\begin{cases} 4x + y = 6 \\ y = -4x - 2 \end{cases}$ $m = -4$
$m = -4$

(parallel)   perpendicular   neither

slope:
$\begin{cases} y = 5x + 3 \\ x + 4y = 8 \end{cases}$ $m = 5$
$m = -\frac{1}{4}$

parallel   perpendicular   (neither)

@iteachalgebra

ANSWER KEY

SOLVING MULTI-STEP EQUATIONS

Solve each equation. Simplify your answer.

$3(x + 4) = 2.5(x - 6)$
$3x + 12 = 2.5x - 15$
$\underline{-2.5x \qquad -2.5x}$
$0.5x + 12 = \quad -15$
$\underline{\quad -12 = \quad -12}$
$\dfrac{0.5x}{0.5} = \dfrac{-27}{0.5}$
$\boxed{x = -54}$

$2(x - 5) + 7 = -3(2x - 6)$
$2x - 10 + 7 = -6x + 18$
$2x - 3 = -6x + 18$
$\underline{+ 6x \qquad + 6x}$
$8x - 3 = \quad 18$
$\underline{\quad + 3 \qquad + 3}$
$\dfrac{8x}{8} = \dfrac{21}{8}$ $\boxed{x = 2.625}$

$\frac{1}{2}(4x - 8) = \frac{3}{4}(8x + 4)$
$2x - 4 = 6x + 3$
$\underline{-2x \qquad -2x}$
$\quad -4 = 4x + 3$
$\underline{\quad -3 \qquad -3}$
$\dfrac{-7}{4} = \dfrac{4x}{4}$
$\boxed{-1.75 = x}$

$\frac{1}{2}x + 5 = \frac{2}{5}x - 8$
$10(\frac{1}{2}x + 5) = 10(\frac{2}{5}x - 8)$
$5x + 50 = 4x - 80$
$\underline{-4x \qquad -4x}$
$x + 50 = \quad -80$
$\underline{\quad -50 \qquad -50}$
$\boxed{x = -130}$

$\frac{2}{3}(5x + 6) = \frac{3}{2}(8x - 4)$
$6[\frac{2}{3}(5x + 6)] = 6[\frac{3}{2}(8x - 4)]$
$4(5x + 6) = 9(8x - 4)$
$20x + 24 = 72x - 36$
$\underline{-72x \qquad -72x}$
$-52x + 24 = \quad -36$
$\underline{\quad -24 = \quad -24}$
$\dfrac{-52x}{-52} = \dfrac{-60}{-52}$
$\boxed{x = \frac{15}{13}}$

$\frac{1}{3}x + \frac{1}{4} = \frac{2}{3}x - \frac{1}{6}$
$12(\frac{1}{3}x + \frac{1}{4}) = 12(\frac{2}{3}x - \frac{1}{6})$
$4x + 3 = 8x - 2$
$\underline{-4x \qquad -4x}$
$\quad 3 = 4x - 2$
$\underline{\quad +2 \qquad +2}$
$\dfrac{5}{4} = \dfrac{4x}{4}$
$\boxed{1.25 = x}$

@iteachalgebra

ANSWER KEY

SOLVING INEQUALITIES

Solve the inequalities.

$30 + 2x < 17$
$\underline{-30 \qquad -30}$
$\dfrac{2x}{2} < \dfrac{-13}{2}$
$\boxed{x < -6.5}$

$15 < -4x + 18$
$\underline{-18 \qquad -18}$
$\dfrac{-3}{-4} < \dfrac{-4x}{-4}$
$\dfrac{3}{4} > x$ $\boxed{x < \dfrac{3}{4}}$

$6 \leq 4x + 80$
$\underline{-80 \qquad -80}$
$\dfrac{-74}{4} \leq \dfrac{4x}{4}$
$-19 \leq x$ $\boxed{x \geq -19}$

$10 - 2x \leq 17$
$\underline{-10 \qquad -10}$
$\dfrac{-2x}{-2} \leq \dfrac{7}{-2}$
$\boxed{x \geq -3.5}$

$-12 > -3x - 12$
$\underline{+12 \qquad +12}$
$\dfrac{0}{-3} > \dfrac{-3x}{-3}$
$0 < x$ $\boxed{x > 0}$

$-9 \leq -5x - 33$
$\underline{+33 \qquad +33}$
$\dfrac{24}{-5} \leq \dfrac{-5x}{-5}$
$-4.8 \geq x$ $\boxed{x \leq -4.8}$

$8 + 2x < -x + 17$
$\underline{-8 \qquad\qquad -8}$
$2x < -x + 9$
$\underline{+x \quad +x}$
$3x < \qquad 9$
$\boxed{x < 3}$

$4x - 9 \leq 5x + 80$
$\underline{-4x \qquad -4x}$
$-9 \leq x + 80$
$\underline{-80 \qquad -80}$
$-89 \leq x$
$\boxed{x \geq -89}$

$5 - 2x \geq 6(x - 3)$
$5 - 2x \geq 6x - 18$
$\underline{-5 \qquad\qquad -5}$
$-2x \geq 6x - 23$
$\underline{-6x \quad -6x}$
$\dfrac{-8x}{-8} \geq \dfrac{-23}{-8}$
$\boxed{x \leq 2.875}$

$-3(3 + x) \leq -6x - 11$
$-9 - 3x \leq -6x - 11$
$\underline{+9 \qquad\qquad +9}$
$-3x \leq -6x - 2$
$\underline{+6x \quad +6x}$
$\dfrac{3x}{3} \leq \dfrac{-2}{3}$
$\boxed{x \leq -\dfrac{2}{3}}$

@iteachalgebra

ANSWER KEY

SUBSTITUTION TO SOLVE SYSTEMS

Solve each system by substitution.

$\begin{cases} y = -2x \\ y = x + 3 \end{cases}$

$-2x = x + 3$
$\underline{-x \qquad -x}$
$-3x = \quad 3$

$\boxed{x = -1}$

$y = -2x$
$y = -2(-1)$
$\boxed{y = 2}$

$\boxed{(-1, 2)}$

$\begin{cases} y = 3x + 3 \\ x - 3y = 9 \end{cases}$

$x - 3(3x + 3) = 9$
$x - 9x - 9 = 9$
$-8x - 9 = 9$
$-8x = 18$
$\boxed{x = -2.25}$

$y = 3x + 3$
$y = 3(-2.25) + 3$
$y = -6.75 + 3$
$\boxed{y = -3.75}$

$\boxed{(-2.25, -3.75)}$

$\begin{cases} 2x + y = 6 \\ x = 2y - 1 \end{cases}$

$2(2y - 1) + y = 6$
$4y - 2 + y = 6$
$5y = 8$
$\boxed{y = 1.6}$

$x = 2y - 1$
$x = 2(1.6) - 1$
$x = 3.2 - 1$
$\boxed{x = 2.2}$

$\boxed{(2.2, 1.6)}$

$\begin{cases} y = \frac{2}{5}x + 3 \\ 2x - 5y = 10 \end{cases}$

$2x - 5(\frac{2}{5}x + 3) = 10$
$2x - 2x + 15 = 10$
$15 \neq 10$

$\boxed{\text{no solution}}$

$\begin{cases} x = -4 \\ y = 5 \end{cases}$

$\boxed{(-4, 5)}$

$\begin{cases} 2x + 3y = 6 \\ y = -3x - 1 \end{cases}$

$2x + 3(-3x - 1) = 6$
$2x - 9x - 3 = 6$
$-7x - 3 = 6$
$-7x = 9$
$\boxed{x = -\frac{9}{7}}$

$y = -3x - 1$
$y = -3(-\frac{9}{7}) - 1$
$y = \frac{27}{7} - 1$
$\boxed{y = \frac{20}{7}}$

$\boxed{(-\frac{9}{7}, \frac{20}{7})}$

@iteachalgebra

ANSWER KEY

FACTORING TRINOMIALS

Factor each trinomial.

$x^2 + 5x + 4$

$(x + 1)(x + 4)$

$x^2 + 8x + 16$

$(x + 4)(x + 4)$
$(x + 4)^2$

$x^2 - 6x + 8$

$(x - 2)(x - 4)$

$x^2 - 6x - 7$

$(x + 1)(x - 7)$

$x^2 + 5x + 6$

$(x + 2)(x + 3)$

$x^2 - 10x + 25$

$(x - 5)(x - 5)$
$(x - 5)^2$

$2x^2 + 7x + 3$

$(2x + 1)(x + 3)$

$3x^2 - 13x + 4$

$(3x - 1)(x - 4)$

$5x^2 + 7x - 6$

$(5x - 3)(x + 2)$

Solve the polynomial equation.

$x^2 + 9x = -8$
$x^2 + 9x + 8 = 0$
$(x + 1)(x + 8) = 0$

$x + 1 = 0 \quad x + 8 = 0$
$\underline{-1 \; -1} \quad \underline{-8 \; -8}$
$x \quad = -1 \quad x \quad = -8$

$x = \{-8, -1\}$

$2x^2 = 7x - 3$
$2x^2 - 7x + 3 = 0$
$(2x - 1)(x - 3) = 0$

$2x - 1 = 0 \quad x - 3 = 0$
$\underline{+1 \; +1} \quad \underline{+3 \; +3}$
$\dfrac{2x}{2} = \dfrac{1}{2} \quad x = 3$

$x = \{\tfrac{1}{2}, 3\}$

$3x^2 + 15x = -18$
$3x^2 + 15x + 18 = 0$
$3(x^2 + 5x + 6) = 0$
$3(x + 2)(x + 3) = 0$

$x + 2 = 0 \quad x + 3 = 0$
$\underline{-2 \; -2} \quad \underline{-3 \; -3}$
$x \quad = -2 \quad x \quad = -3$

$x = \{-3, -2\}$

@iteachalgebra

ANSWER KEY

SIMPLIFYING RADICALS

Simplify each radical expression.

$\sqrt{4}$
2

$\sqrt{6}$
$\sqrt{6}$

$\sqrt{8}$
$\sqrt{4}\sqrt{2}$
$2\sqrt{2}$

$\sqrt{9}$
3

$\sqrt{10}$
$\sqrt{10}$

$\sqrt{12}$
$\sqrt{4}\sqrt{3}$
$2\sqrt{3}$

$\sqrt{18}$
$\sqrt{9}\sqrt{2}$
$3\sqrt{2}$

$\sqrt{25}$
5

$\sqrt{28}$
$\sqrt{4}\sqrt{7}$
$2\sqrt{7}$

$\sqrt{32}$
$\sqrt{16}\sqrt{2}$
$4\sqrt{2}$

$\sqrt{40}$
$\sqrt{4}\sqrt{10}$
$2\sqrt{10}$

$\sqrt{48}$
$\sqrt{16}\sqrt{3}$
$4\sqrt{3}$

$\sqrt{50}$
$\sqrt{25}\sqrt{2}$
$5\sqrt{2}$

$\sqrt{55}$
$\sqrt{55}$

$\sqrt{60}$
$\sqrt{4}\sqrt{15}$
$2\sqrt{15}$

$\sqrt{64}$
8

$\sqrt{72}$
$\sqrt{36}\sqrt{2}$
$6\sqrt{2}$

$\sqrt{90}$
$\sqrt{9}\sqrt{10}$
$3\sqrt{10}$

$\sqrt{99}$
$\sqrt{9}\sqrt{11}$
$3\sqrt{11}$

$\sqrt{120}$
$\sqrt{4}\sqrt{30}$
$2\sqrt{30}$

$\sqrt{150}$
$\sqrt{25}\sqrt{6}$
$5\sqrt{6}$

$\sqrt{160}$
$\sqrt{16}\sqrt{10}$
$4\sqrt{10}$

$\sqrt{200}$
$\sqrt{100}\sqrt{2}$
$10\sqrt{2}$

$\sqrt{256}$
16

$\sqrt{300}$
$\sqrt{100}\sqrt{3}$
$10\sqrt{3}$

@iteachalgebra

ANSWER KEY

OPERATIONS WITH RADICALS

10

Simplify each radical expression.

$\sqrt{2} + \sqrt{2}$
$2\sqrt{2}$

$4\sqrt{3} + \sqrt{3}$
$5\sqrt{3}$

$5\sqrt{6} + 2\sqrt{6}$
$7\sqrt{6}$

$\sqrt{2} - \sqrt{2}$
0

$4\sqrt{3} - \sqrt{3}$
$3\sqrt{3}$

$5\sqrt{6} - 2\sqrt{6}$
$3\sqrt{6}$

$\sqrt{2} \cdot \sqrt{2}$
$\sqrt{4} = 2$

$4\sqrt{3} \cdot \sqrt{3}$
$4\sqrt{9} = 4(3)$
$= 12$

$5\sqrt{6} \cdot 2\sqrt{6}$
$10\sqrt{36} = 10(6)$
$= 60$

$\sqrt{72} + \sqrt{50}$
$\sqrt{36}\sqrt{2} + \sqrt{25}\sqrt{2}$
$6\sqrt{2} + 5\sqrt{2}$
$11\sqrt{2}$

$4\sqrt{45} - \sqrt{125}$
$4\sqrt{9}\sqrt{5} - \sqrt{25}\sqrt{5}$
$4(3)\sqrt{5} - 5\sqrt{5}$
$12\sqrt{5} - 5\sqrt{5}$
$7\sqrt{5}$

$5\sqrt{27} + 2\sqrt{5}$
$5\sqrt{9}\sqrt{3} + 2\sqrt{5}$
$5(3)\sqrt{3} + 2\sqrt{5}$
$15\sqrt{3} + 2\sqrt{5}$

@iteachalgebra

CLASSIFYING SEGMENTS, RAYS, & LINES

Determine the segments, rays, and lines from the diagram.

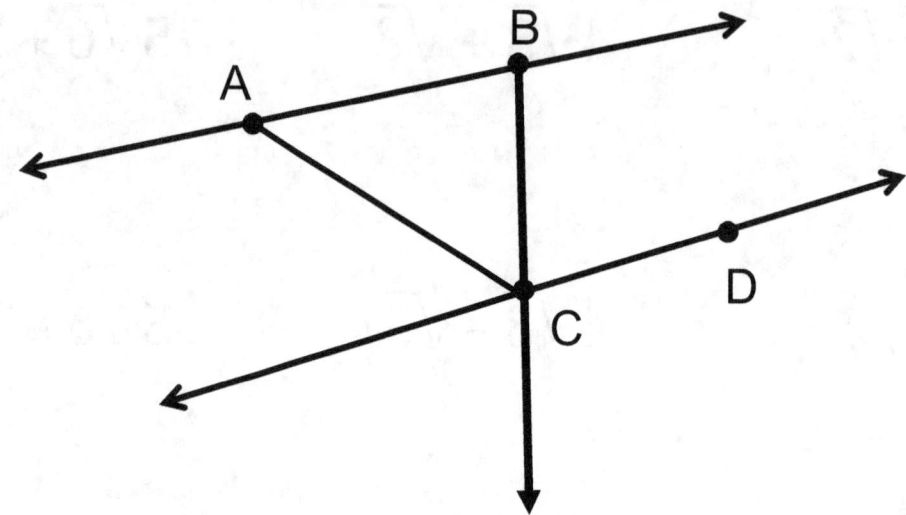

SEGMENTS	RAYS	LINES
$\overline{AC}$	$\overrightarrow{BC}$	$\overleftrightarrow{AB}$
		$\overleftrightarrow{CD}$

Determine whether each statement is true or false.

Two lines can intersect at exactly one point.	true
Two lines can intersect at exactly two points.	false
The are an infinite number of points on a line.	true
A ray has an arrow at one end.	true
A segment and a line are identical.	false

@iteachalgebra

ANSWER KEY   12

NUMBER OF EDGES & VERTICES

List the number of edges and vertices for each figure.

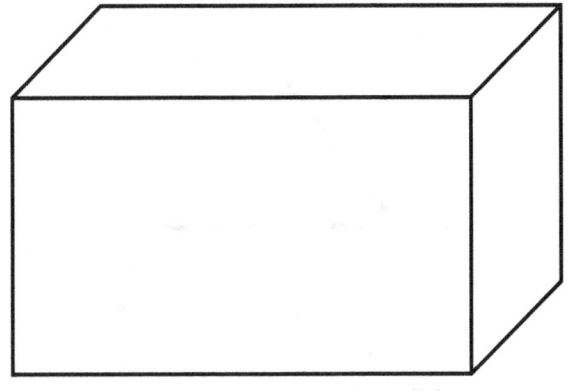

rectangular prism

edges: 12

vertices: 8

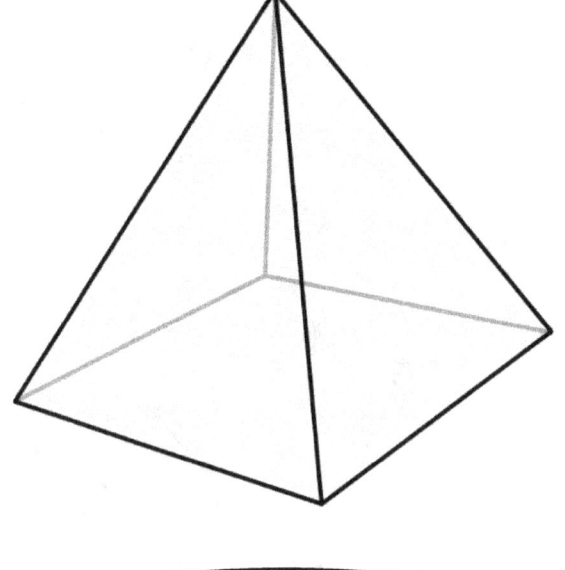

square pyramid

edges: 8

vertices: 5

cylinder

edges: none

vertices: none

@iteachalgebra

ANSWER KEY

ANGLE MEASUREMENTS

13

Circle the type of angle shown and the best approximate measure of the angle.

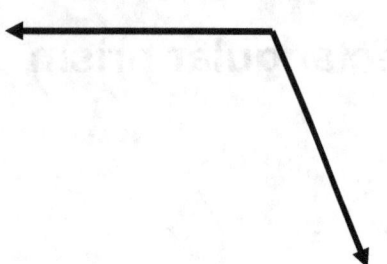

acute 60
(obtuse) (100)
right 90

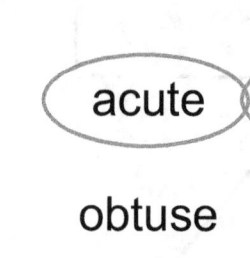

(acute) (60)
obtuse 100
right 90

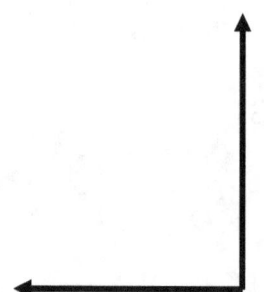

acute 60
obtuse 100
(right) (90)

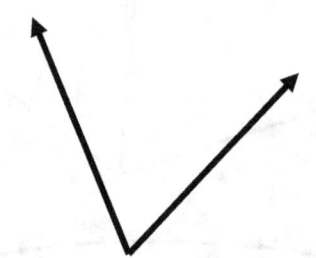

(acute) (60)
obtuse 100
right 90

@iteachalgebra

ANSWER KEY

TYPES OF TRIANGLES

Name the triangle based on its sides and angles.
Names include equilateral, isosceles, and scalene, acute, obtuse, and right.

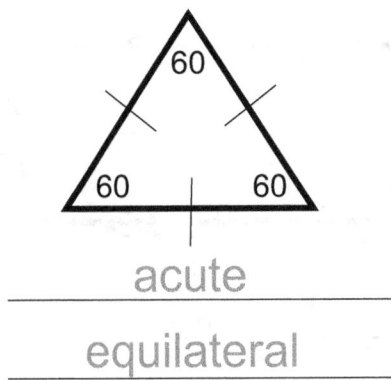

acute

equilateral

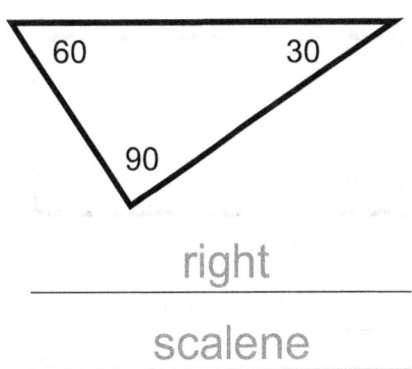

right

scalene

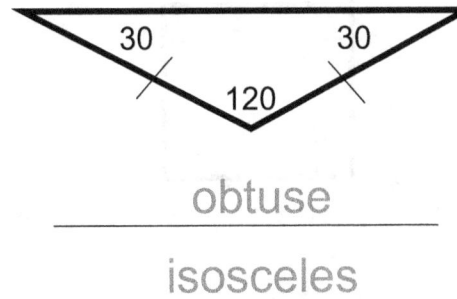

obtuse

isosceles

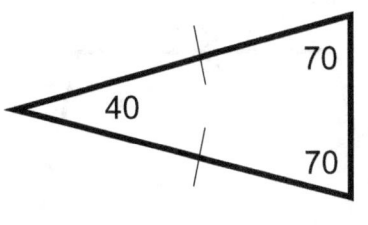

acute

isosceles

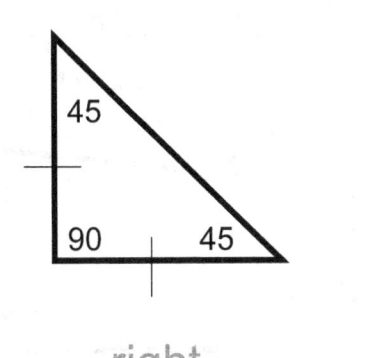

right

isosceles

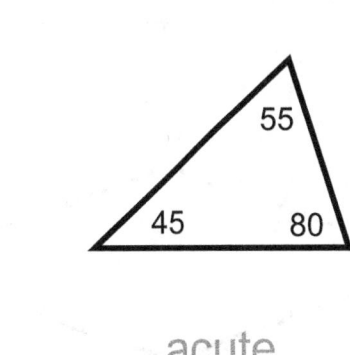

acute

scalene

@iteachalgebra

ANSWER KEY
TYPES OF QUADRILATERALS

Determine if the quadrilateral is a square, rectangle, rhombus, trapezoid, isosceles trapezoid, parallelogram, or more than one of those names.

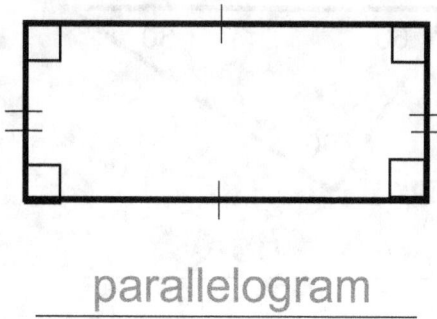

parallelogram

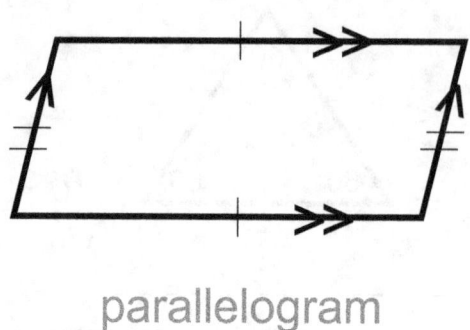

parallelogram

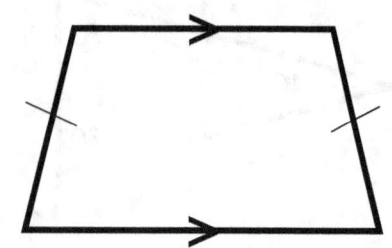

trapezoid

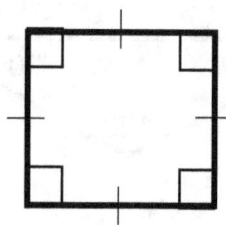

parallelogram

rectangle, square

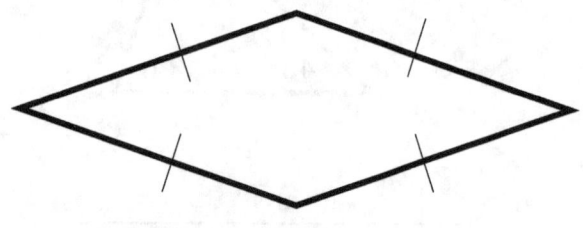

parallelogram

rhombus

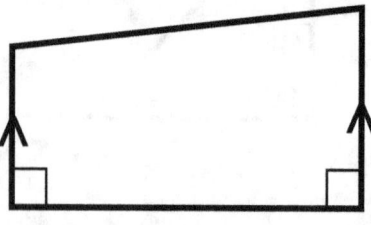

trapezoid

@iteachalgebra

ANSWER KEY

PARTS OF A CIRCLE

16

Given the circle, name each part.

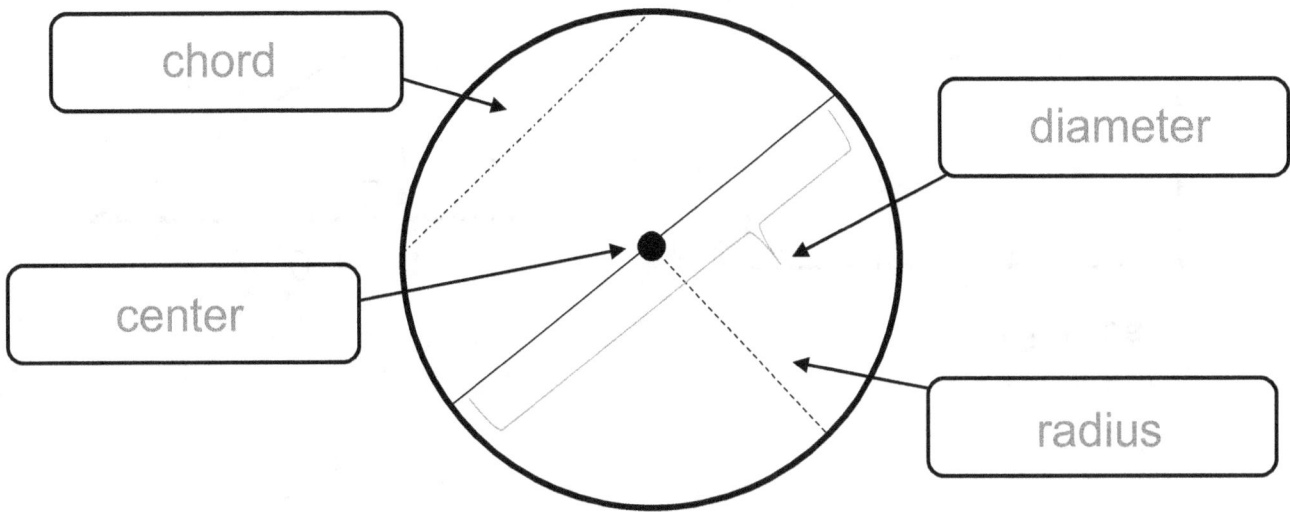

chord

diameter

center

radius

Find the circumference and area of each circle.

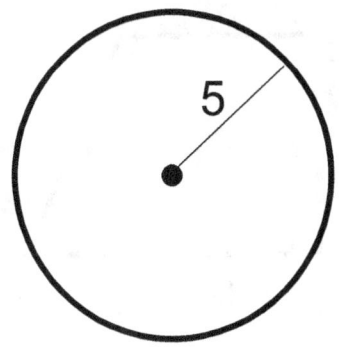

Circumference: $C = 2\pi r$

$C = 2\pi r$

$C = 2\pi(5)$

$C = 10\pi$

$C \sim 31.4$ units

Area: $A = \pi r^2$

$A = \pi r^2$

$A = \pi(5)^2$

$A = 25\pi$

$A \sim 78.5$ units

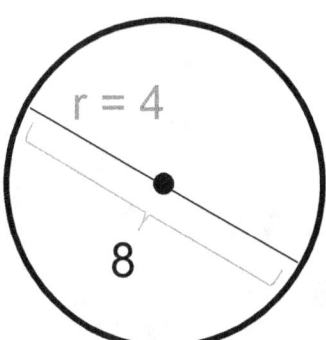

Circumference: $C = \pi d$

$C = \pi d$

$C = \pi(8)$

$C = 8\pi$

$C \sim 25.1$ units

Area: $A = \pi r^2$

$A = \pi r^2$

$A = \pi(4)^2$

$A = 16\pi$

$A \sim 50.3$ units

@iteachalgebra

ANSWER KEY   17

AREA FORMULAS

Calculate the area of each figure.

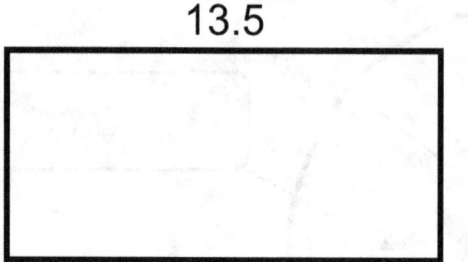

rectangle
A = lw
A = (13.5)(4)
A = 54 units²

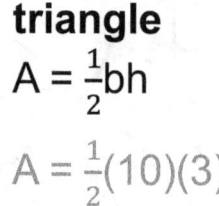

triangle
$A = \frac{1}{2}bh$
$A = \frac{1}{2}(10)(3)$
A = 15 units²

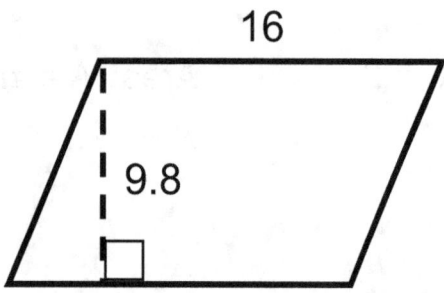

parallelogram
A = bh
A = (16)(9.8)
A = 156.8 units²

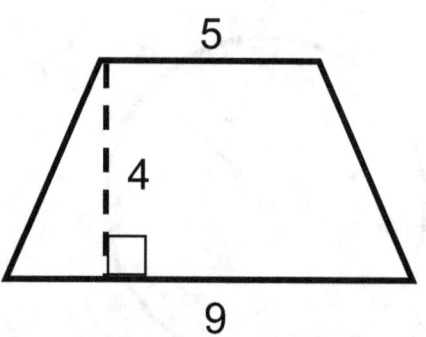

trapezoid
$A = \frac{1}{2}h(b_1 + b_2)$
$A = \frac{1}{2}(4)(5 + 9)$
A = 2(14)
A = 28 units²

@iteachalgebra

ANSWER KEY   18

VOLUME FORMULAS

Calculate the volume of each figure.

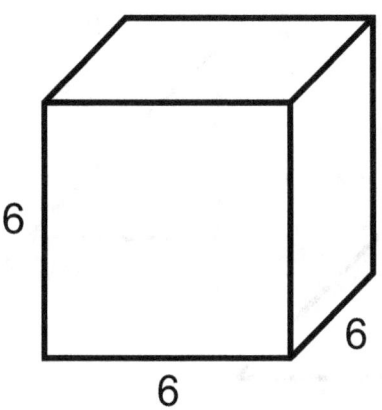

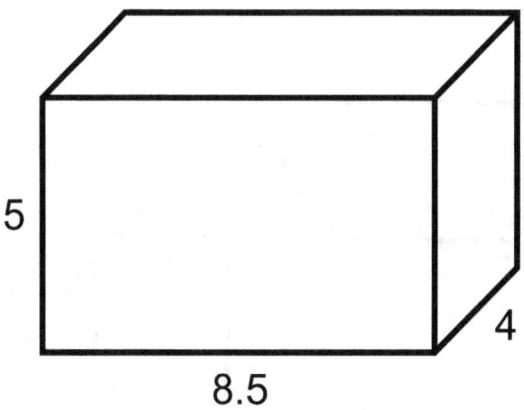

cube
V = s^3
V = $(6)^3$
V = 216 units^3

rectangular prism
V = lwh
V = (8.5)(4)(5)
V = 170 units^3

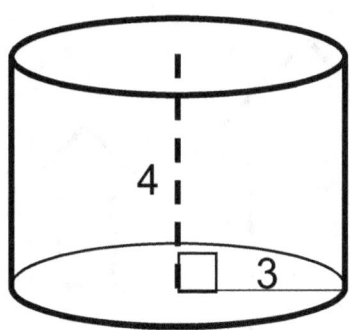

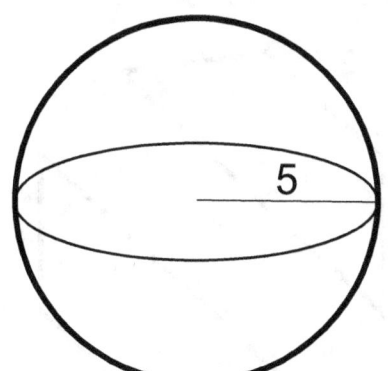

cylinder
V = $\pi r^2 h$
V = $\pi(3)^2(4)$
V = $\pi(9)(4)$
V = 36π units^3
V ~ 113.1π units^3

sphere
V = $\frac{4}{3}\pi r^3$
V = $\frac{4}{3}\pi(5)^3$
V = $\frac{4}{3}\pi(125)$
V = $\frac{500}{3}\pi$ units^3
V ~ 523.6 units^3

@iteachalgebra

ANSWER KEY

TRANSFORMATIONS

19

Determine the type of transformation shown in each diagram as a translation, rotation, reflection, or dilation.

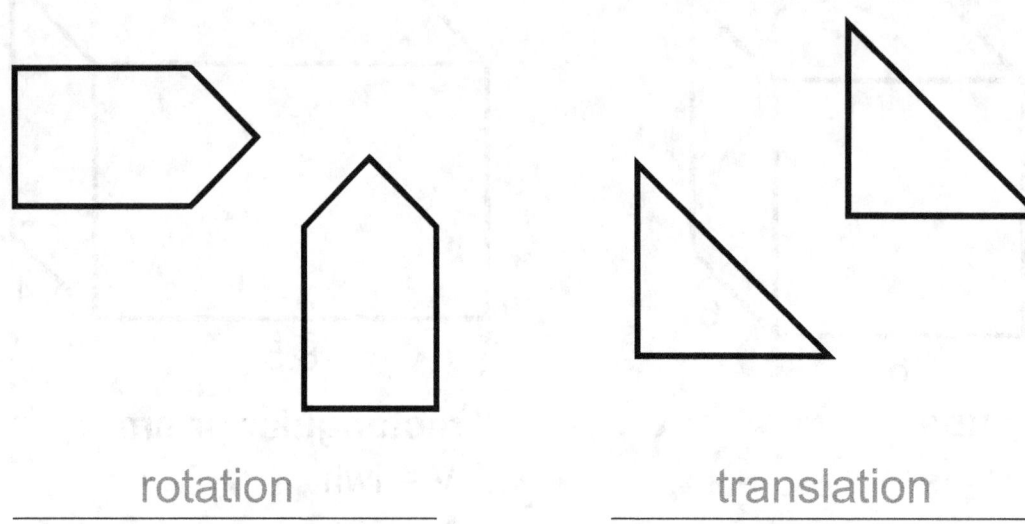

rotation

translation

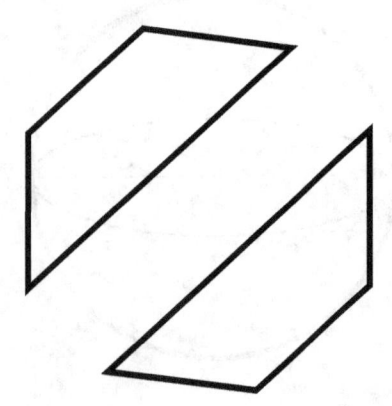

reflection or rotation

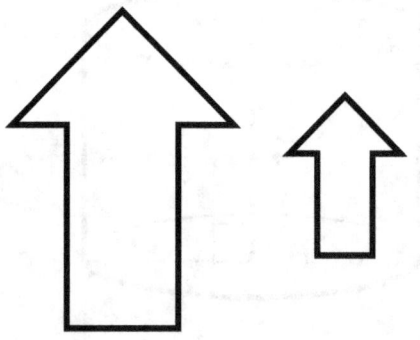

dilation

@iteachalgebra

ANSWER KEY

CONGRUENT OR SIMILAR

Determine whether the figures shown are congruent or similar.

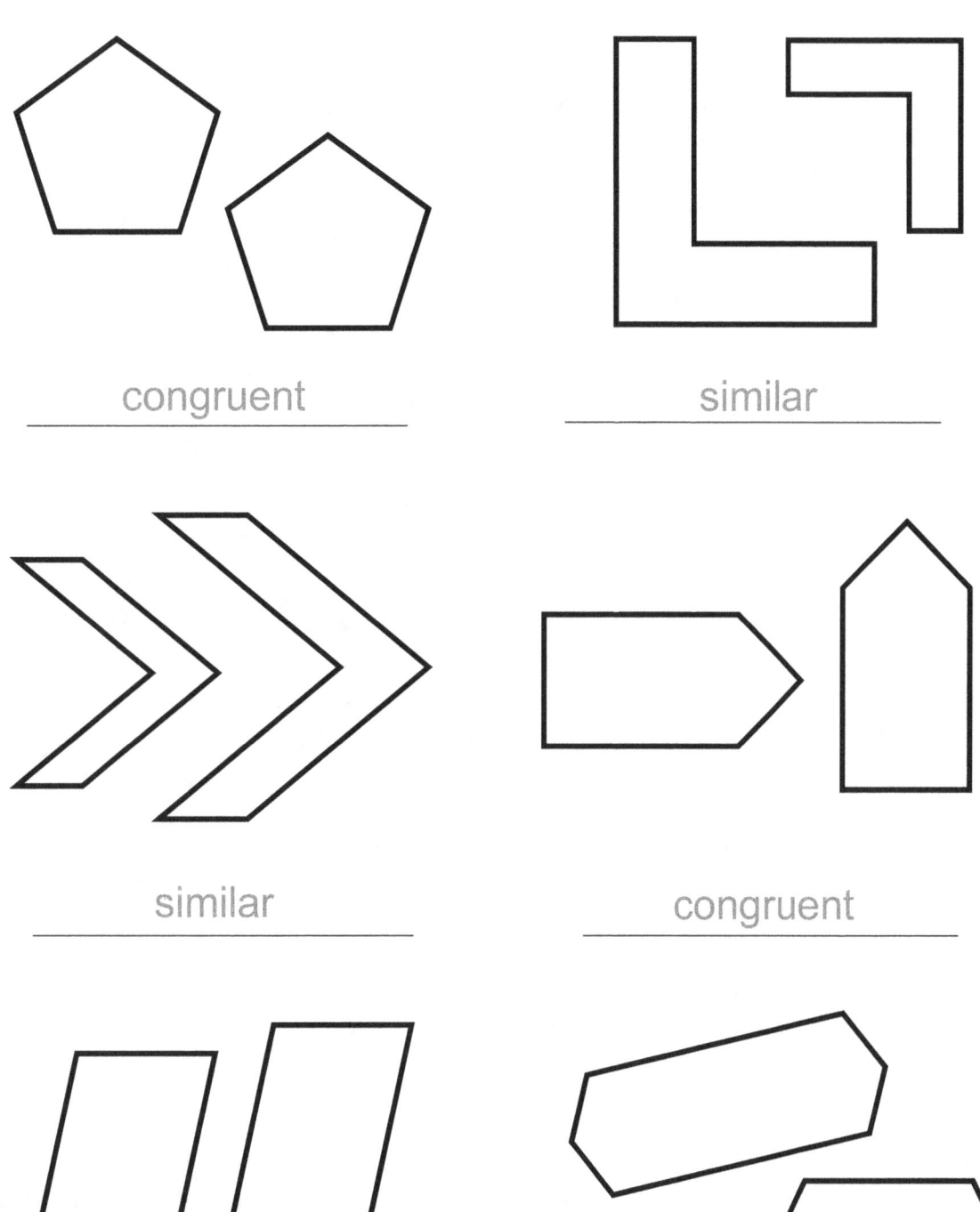

congruent

similar

similar

congruent

congruent

similar

@iteachalgebra